Impedance matching using double stubs

Ain Rehman

DEDICATION

I dedicate this book to my family, my teachers and all those whose help was so invaluable to me through the years, in person and online

Double stub matching technique: Single stub tuning has advantages and disadvantages. It is simple to understand and use. However, if the load impedance changes then we have to adjust 'd' every time. This can be quite tedious. In double stub tuning (Figure M.14) the distance between stubs and 'd' can be made fixed and only the stub lengths changed to tune a wide band of loads. However, even double stub tuning has the disadvantage of *not* being able to match all impedances and we must resort to triple stub tuning. In this book we examine double stub tuning to understand what it is and how it is done. The advantages and disadvantages will be made clear as we work through this technique of impedance matching.

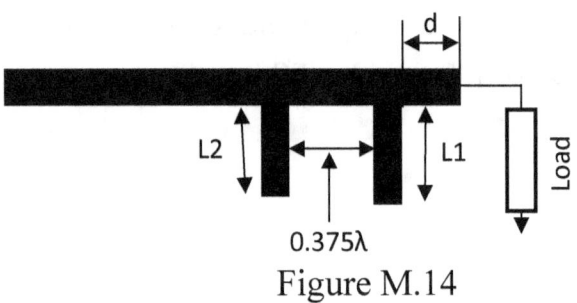

Figure M.14

Double stub matching: Double stub impedance matching is more involved than single stub matching. First of all we need to discuss the rationale for using double stub matching. In order to understand all this, consider the points below:

Impedance matching using double stubs

1.0   A single stub match involves the use of a stub of fixed length placed at a fixed position on the transmission line from the load for a specific load.

2.0   If the value of the load changes then the length of the stub and the position at which it is placed must also change.

3.0   A better technique would be where we could fix the positions of the stubs in relation to each other and only change the lengths of the stubs to match varying loads. This is done using the techniques of double and triple stub matching.

4.0   Both analytic and graphical techniques are available to do double stub matching. The graphical method is usually the Smith Chart method and that is what we will focus on in this discussion.

5.0   In order to further understand the technique of double stub matching we will briefly digress to discuss some aspects of the Smith Chart that will clarify the technique for readers.

6.0   There are some loads that cannot be matched using double stub matching by simply altering the length of the stubs. However if we are willing and able, to move the stubs together a distance away (or towards the load) then we can accommodate the load. This will be discussed and illustrated using the Smith Chart method. These loads form an area of the Smith Chart collectively known as the *forbidden zone* for double stub matching. For example, if the stub spacing is $\lambda/8$, $3\lambda/8$ or $5\lambda/8$, then the forbidden zone is the entire area of the Smith Chart encircled by the $g = 2$ circle. If it is $\lambda/4$ then the forbidden zone is the area surrounded by the $g = 1$ circle.

7.0 Triple stub matching can be used to overcome the limitations of double stub matching if needed. Similar techniques to that of double stub matching are used for triple stub matching. However triple stub matching is not commonly used in microstrip circuits.

Background for the use of double stub matching:

8.0 *The microstrip (or transmission line) as a transformer.* A length of microstrip with a characteristic impedance Zo acts as a transformer. To visualize this, we can look at the following Smith Chart plot that shows this. In addition, note that the Smith Chart circumference is a measure of the distance moved on a transmission line. For one complete revolution we travel half a wavelength on the line. From the chart its easy to see that a movement from point 1 to point 2 changes the imaginary component (reactance or susceptance) of the line. This is a transformation. So as we move on a transmission line, the line acts as a transformer. This fact is used very successfully in the design of the popular quarter wave transformer.

Note that one complete rotation around the Smith Chart is one half wavelength on the line. This is also represented by 360 degrees around the Smith Chart. Also movement on a uniform transmission line also implies movement on the VSWR circle.

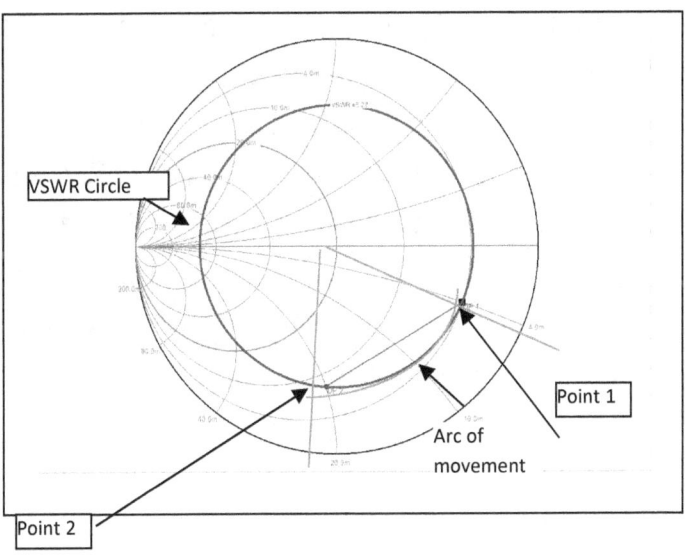

9.0   *The spacing circle.* In double stub matching the two stubs are spaced a predetermined distance away from each other. These distances are typically $\lambda/8$, $\lambda/4$, $3\lambda/8$, $5\lambda/8$ etc. Knowing what we know about a length of line acting as a transformer, we know that the length of line between the two stubs acts as a transformer. *The action of this transformer is to convert the admittance at the position of stub 2 to a different admittance at the **position of stub 1**. So we start from the position of stub 2.*

In order that stub 2 can be finally used to match the line admittance, the real part of the admittance at the position of stub 2, on the line, has to be 1.0 (normalized value). Its susceptance is then jB. jB is the susceptance that is cancelled using stub 2 to ultimately get the matching to the line admittance. The admittance at the position of stub 2, (without the stub) lies on the constant conductance, g = 1 circle. The admittances on the g = 1 circle are all the possible admittances at the stub 2 position for a match to take place.

To reiterate, as a result of these deliberations, that some point of the VSWR circle formed by the position of stub 2, *must* intersect the g=1 or the unity conductance circle on the Smith   Chart.

We also conclude, that the admittance at the position of the first stub, <u>must lie on a circle of</u> <u>*equal radius*</u> <u>but having its center rotated (moved</u> <u>to, or displaced) by the spacing between the stubs</u> <u>*towards the load. Lets call this 'x'.*</u>

This circle is called the *spacing circle* and its construction is explained further below. The concept of the spacing circle is important in understanding the graphical method of double stub matching using the Smith Chart. [Note that *one complete* *rotation around the Smith Chart is <u>360</u>* *<u>Degrees</u> for λ/2*. This fact allows us to graphically construct movements on the line on a Smith Chart. For example a λ/4 length of line is a movement of 180 Degrees on the Smith Chart.]

Consider the Figure shown below where we have assumed a stub spacing of λ/4.

Impedance matching using double stubs

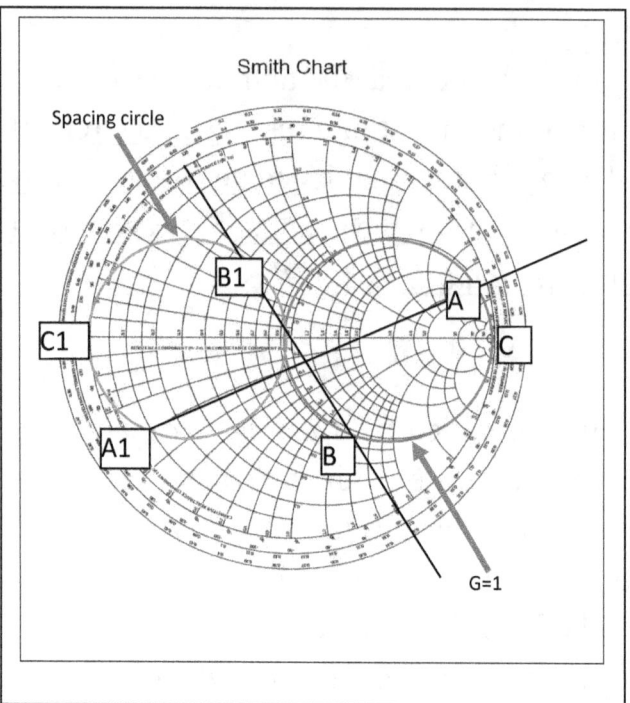

Note that a spacing of λ/4 implies a rotation or displacement by 180 Degrees.

Note: Remember that moving on a transmission line (uniform) is reflected by moving on the VSWR circle on the Smith Chart

In this figure points A, B and C have been "rotated" or moved by 180 degrees to new positions A1, B1, C1. In the same way *all points* on the first circle on the right can be moved to the circle on left thereby accomplishing *the rotation* of the circle by $\lambda/4$. This also shows how the impedances (or admittances as the case may be) change with distance of movement on the transmission line between the two stubs. A *transformative property* of the transmission line.

So it can be seen, that once we know the distance between stubs, the admittances on the line can be transformed by a movement or *spacing,* **towards the load** as shown. The admittances lie on a circle of the same radius but displaced *towards the load*. *Any* spacing between stubs can be handled in this way and spacing circles formed. The center of the spacing circle has been rotated by the length of line between stubs. The radius is still the same only the center has moved to a new position dictated by the distance between the stubs.

Similar transformations can be made for other values of x. In every case the locus of points on the unit conductance circle maps into a new *spacing circle* of the same radius, whose center lies *x* wavelengths *towards the load* (counter clockwise) from the center of the unit conductance circle.

Any intersections of the original VSWR circle with the g=1 circle also are transformed by this procedure and now lie on the spacing circle. In this way we see how the length of line between the two stubs affects the admittances.

To reiterate: To develop a spacing circle simply move the reference circle that lies at a conductance of g=1 and move the points on it by the angular distance between the two stubs to form the *spacing circle. The spacing circle is also a unity circle.*

The importance of this technique in the double stub matching process cannot be overemphasized. It forms the crux of the technique and should be understood to an *intuitive extent* by the engineer who is interested in graphical methods of stub tuning using the Smith Chart.

A further important concept in double stub matching is the *forbidden region*. This is a region on the Smith Chart which consists of impedances that cannot be matched. In his original work Smith describes these regions based on some discrete stub spacings.

The analytical treatment of forbidden regions is somewhat involved but to state the results let us define *gload* as the conductance of the load. Then the forbidden region may be defined by the following constraint:

$$0 \leq \text{gload} \leq 1/\sin^2\beta x$$

where ,

$$x = \text{spacing between stubs}$$

$$\beta = \text{wave number} = 2\pi/\lambda$$

i.e The forbidden region is surrounded by a constant conductance circle whose value depends on the electrical stub separation, x / λ. For example, lets assume that the separation between the stubs,

$$x = 3\lambda/4$$

so,

$$\beta x = 3\pi/4$$

or,

$$1/\sin^2\beta x = 2.0$$

Its clear from this, that the forbidden region is enclosed by the gload = 2 circle on the Smith Chart.

These identities can be used to define the forbidden regions on the Smith Chart for various loads.

Modification of double stub matching to override the forbidden region restriction:

If we find that the load admittance lies in a forbidden region we can still use double stub matching if we insert a length of line between the load and the first stub of length = *lenz.*

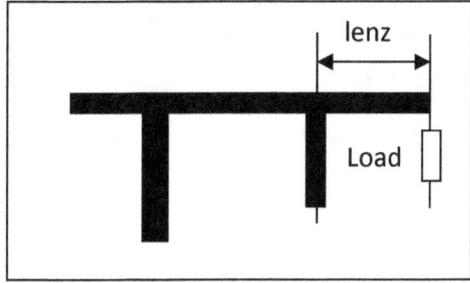

The effect of the added length is to change the load admittance so that the transformed value of *gLoad* is *transformed* to a level that meets the requirements of double stub matching. See the figures below.

Impedance matching using double stubs

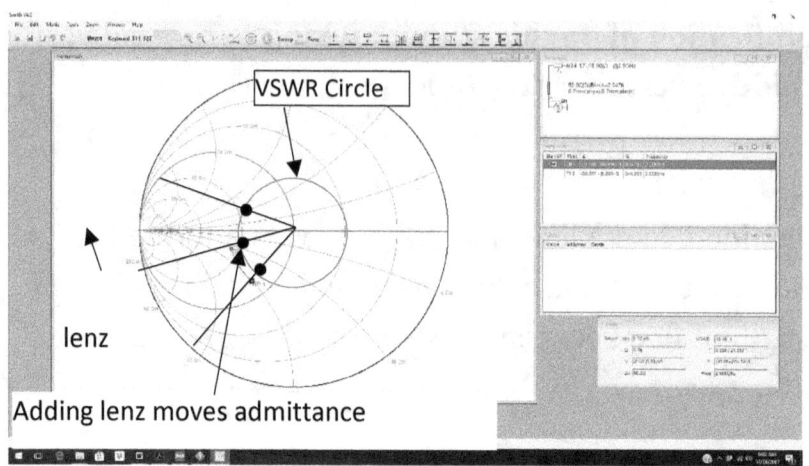

Adding lenz moves admittance

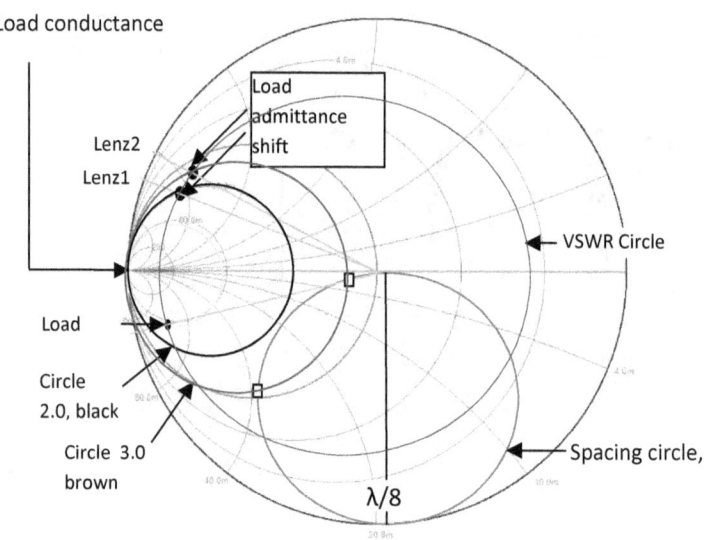

Generate the spacing circle with a shift of λ/8
towards the load. (Admittance Chart)

Points of intersection of the conductance circle (circle 3) with the spacing circle when lenz=lenz2 in yellow (small boxes). Please note that moving on a uniform transmission line is reflected by moving on a VSWR circle on the Smith Chart.

Notes on the two figures above.

1.0 The first figure shows that as a length of microstrip (or transmission line) is added, the admittance moves to a new value on the VSWR circle. A transformative function.

2.0 The second figure is much more involved. We will break it down as shown below:

3.0 An admittance is shown on a Smith Chart along with its conductance circle shown in green (see the arrow for "load"). For a black and white diagram the arrow should distinguish it from other circles.

4.0 It is assumed that the two stubs are spaced by λ/8. As a result the spacing circle is rotated by this amount and is shown in red (along with its caption with an arrow).

5.0 Lengths of line, lenz1 and lenz2 are added sequentially. These line lengths move the admittance values on the VSWR circle shown in brown (or arrow).

6.0 It can be seen quite plainly that the non transformed admittance lies on a conductance circle which _does not_ intersect the rotated spacing circle at any point. As a result the two stubs cannot be used for matching this admittance.

7.0 As the admittance is moved by the addition of lenz1 and lenz2 it is found that the conductance circle for lenz2 intersects the spacing circle at two points. Therefore matching can be done using the double stubs.

8.0 If the first stub _has_ to be right at the position of the load, then the distance _between_ the two stubs has to be changed such that the conductance circle of the load and the spacing circle _does_ intersect.

Applying these concepts to double stub matching:

Lets apply these concepts to double stub matching. Refer to the figure below which shows the two circles. The g = 1 circle and the λ/4 spacing circle to start with.

The blue circle represents stub 2 position without the stub present. The red circle represents the spacing circle for a quarter wavelength spacing of the stubs. It also represents the admittances that are the result of the transformation effected by the length of line say, $x = \lambda/4$.

In the next figure (figure match 1) we place the load admittance on the chart as shown at point P1.

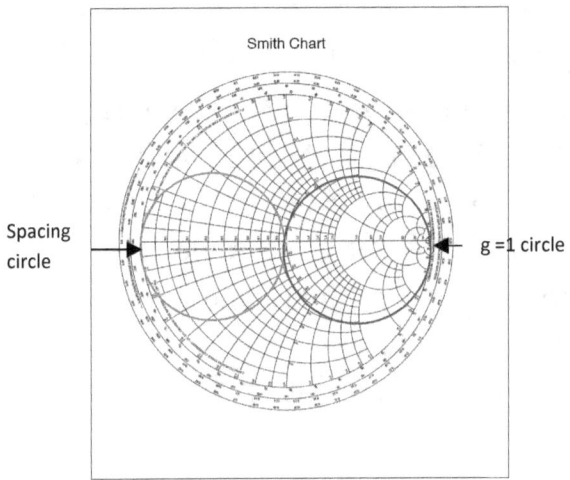

The g = 1 circle (blue) and the spacing circle for quarter wave length of spacing between the stubs

Its interesting to note that the spacing circle and the load conductance circle represent what is happening at the *load end* of the circuit. There are two points of intersection of the load conductance circle with the spacing circle, P2 and P3. We can use either to perform the matching.

In this case let us use P2. To understand what is happening here note that the spacing circle and the load conductance circle represent the action *at the load end* of the circle as stated above. What we want to do is to move the load admittance to fall on the point P2. This point, when looked at *from the stub2 end* (by virtue of the transformation property of the line) will cause it to fall on the g=1 circle as needed by the matching requirement. <u>This act represents the reverse of the spacing circle move. Now we are moving from the spacing circle to the g=1 circle</u>. This is of course very valid; it's a reciprocal relationship as noted in previous discussions above. The next figure and its associated discussion shows how this is done. Θ represents the angular movement of the point P1 to point P2.

This can be accomplished by adding a stub with the correct value at the load. The correct value is the *difference between the susceptance at point P1 and P2 or P3.*

See Figure match 1 below.

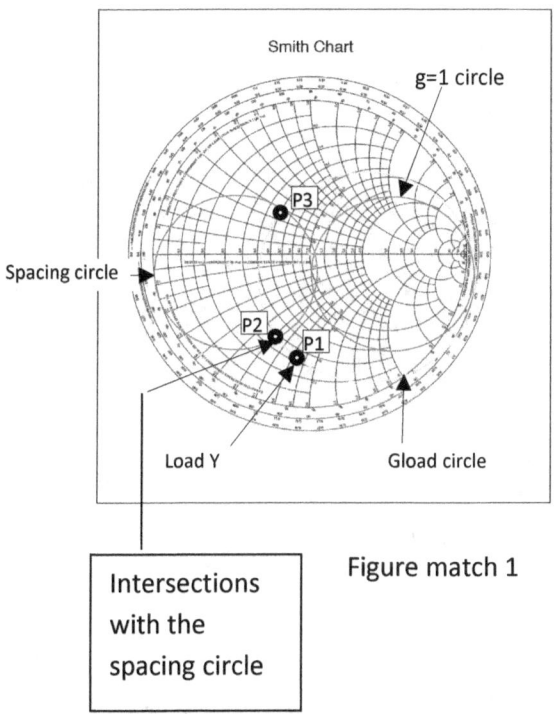

Figure match 1

Once the stub is added to the load side of the line it moves P1 to P2 (or P3) and then the transformation to the stub 2 side shows that the conductance is now unity with a susceptance component. The susceptance component is cancelled using stub 2 with an equal and opposite susceptance.

At this point matching is complete.

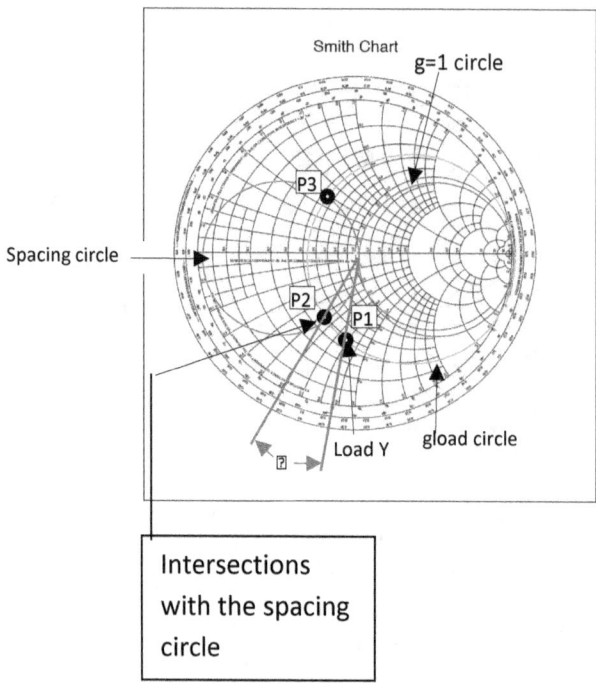

Please note that this book does not address the triple stub matching technique at this time. Triple stub matching can be very useful in a similar way to overcome forbidden regions and thereby match the loads. However, it is not commonly used in microstrip type design.

Double stub matching example.

At this point lets look at an example of double stub matching. The problem can be stated as follows:.

The first stub is placed 0.1λ from the end of the line where the load is situated. The second stub is spaced to the first stub by 0.375λ as shown on the schematic below. The load is also shown.

The frequency of operation is 1 Ghz. The matching circuit is made of microstrip on a 1.7 mm thick FR- 4 PCB, 50 Ohm characteristic impedance is achieved by using a 3 mm line width.

Using these parameters the guide wavelength is calculated as 141 mm.

So the first stub is placed at 14 mm from the load.

The spacing between stubs is 52.875 mm. We have to calculate the length of the stubs.

The final schematic is shown below.

Impedance matching using double stubs

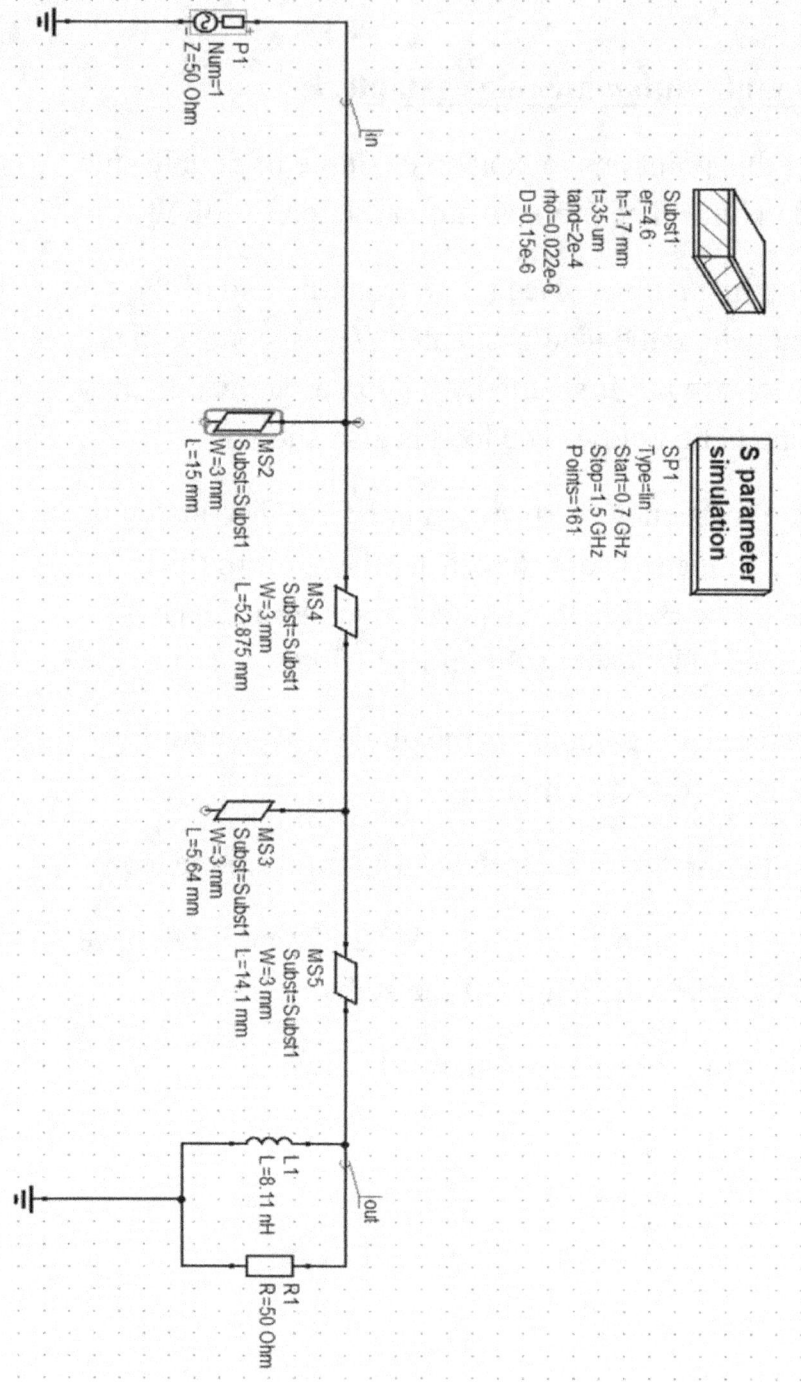

Impedance matching using double stubs

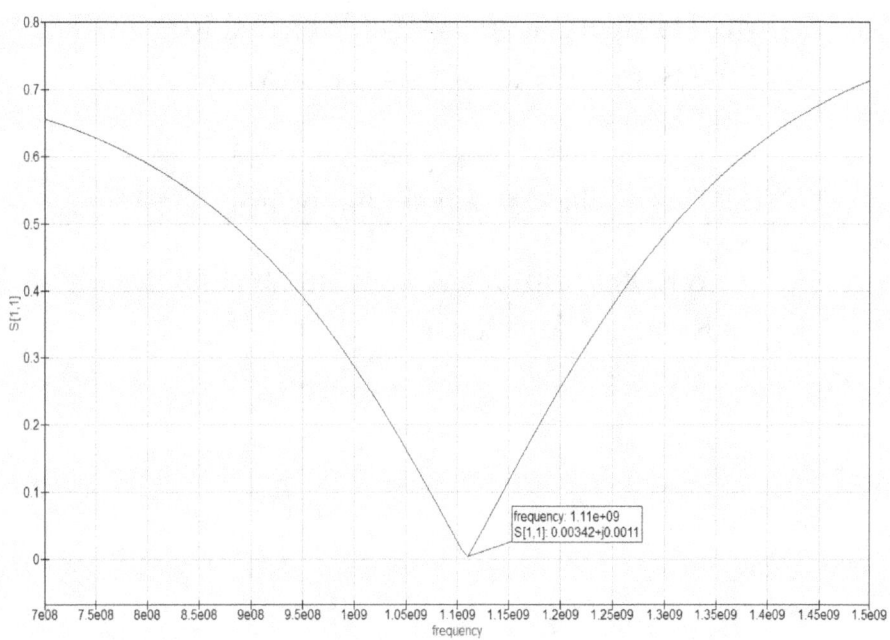

frequency: 1.11e+09
S[1,1]: 0.00342+j0.0011

The input reflection coefficient of the circuit showing degree of match S11 ~ 0.0 at frequency ~ 1 Ghz.

The figure above shows the result of the design using a public domain simulator.

References:

1.0 Smith Chart tool from the web. developed by Prof. Fritz Dellsperger, Juerg Tschirren and Roger Wetzel of the Berne Institute of Engineering and Architecture.

2.0 QUCS a public domain simulator from the Web,

3.0 "VSWR and Impedance matching techniques", By Ain Rehman, Amazon.

www.ingramcontent.com/pod-product-compliance
Lightning Source LLC
Chambersburg PA
CBHW071204220526
45468CB00003B/1155